# LESSONS FROM FOLLOW THE SCIENCE

## Insights And Reflection For Implementing Sharyl Attkison Principles

Andrey Matts

Copyright © 2024 by Andrey Matts

All rights reserved. No part of this book may be reproduced or transmitted in any form or by any means, electronic or mechanical, including photocopying and recording without permission in writing from the copyright owner.

**Disclaimer Note**

The content provided in this book is for informational purposes only. The techniques and strategies discussed are based on personal preference and should not replace the original book.

# Introduction to Big Pharma's Influence

**5 Insights:**

1. Big Pharma has a vast influence on healthcare policies and decisions.
2. The pharmaceutical industry prioritizes profits over patient care.
3. Media, government, and medical professionals can be complicit.
4. Ethical violations are often overlooked.
5. Investigative journalism plays a crucial role in exposing misconduct.

**5 Reflection Questions:**

1. How does Big Pharma influence healthcare policies?

_____

_____

_____

_____

2. What are the consequences of prioritizing profits over patients?

_____

_____

_____

_____

3. In what ways do media and government contribute to pharmaceutical misconduct?

_____

_____

_____

_____

4. Why are ethical violations often ignored in the pharmaceutical industry?

_____

_____

_____

_____

5. How can investigative journalism help address these issues?

_____

_____

_____

_____

# The History of Pharmaceutical Misconduct

**5 Insights:**

1. Pharmaceutical misconduct has a long history of unethical practices.
2. Early scandals set a precedent for future industry behavior.
3. Many cases involve prioritizing profit over patient safety.
4. Legal settlements often do not deter repeated misconduct.
5. Public trust erodes with ongoing industry malpractices.

**5 Reflection Questions:**

1. What are some historical examples of pharmaceutical misconduct?

_____

_____

_____

2. How have early scandals shaped today's pharmaceutical practices?

_____

_____

_____

_____

3. Why does the industry prioritize profit over safety?

_____

_____

_____

_____

4. How effective are legal settlements in preventing future misconduct?

_____

_____

_____

_____

5. How does repeated misconduct affect public trust in the industry?

_____

_____

_____

_____

# Understanding Financial Ties and Conflicts of Interest

**5 Insights:**

1. Financial ties between pharmaceutical companies and healthcare professionals are widespread.
2. These relationships often create conflicts of interest.
3. Such conflicts can compromise patient care and research integrity.
4. Financial incentives can bias medical recommendations.
5. Transparency about these ties is often lacking.

**5 Reflection Questions:**

1. How do financial ties between Big Pharma and healthcare professionals impact patient care?

_____

_____

2. What are the risks of conflicts of interest in medicine?

3. Why might financial incentives bias medical research and recommendations?

4. How can transparency improve trust in healthcare?

5. What measures can reduce conflicts of interest in the pharmaceutical industry?

# Media Collusion and Information Control

**5 Insights:**

1. Big Pharma heavily influences media narratives about drugs and treatments.
2. Media outlets often rely on pharmaceutical advertising revenue.
3. Negative information about drugs is sometimes suppressed.
4. Biased reporting shapes public perception of medical products.
5. Independent journalism is crucial for unbiased information.

**5 Reflection Questions:**

1. How does Big Pharma influence media coverage?

_____

_____

_____

2. Why might media outlets suppress negative drug information?

_____
_____
_____
_____

3. What are the effects of biased reporting on public perception?

_____
_____
_____
_____

4. How does pharmaceutical advertising impact news objectivity?

_____

_____

_____

_____

5. What role does independent journalism play in informing the public?

_____

_____

_____

_____

# Misinformation in Medical Education

**5 Insights:**

1. Pharmaceutical companies often provide biased educational materials to medical schools.
2. These materials can influence doctors' prescribing habits.
3. Biased education impacts patient care and treatment choices.
4. Financial incentives encourage reliance on pharma-sponsored information.
5. Independent, unbiased education is crucial for medical professionals.

**5 Reflection Questions:**

1. How does Big Pharma influence medical education?

---------------------------------

---------------------------------

---------------------------------

2. What impact does biased education have on patient care?

_____
_____
_____
_____

3. Why do doctors rely on pharmaceutical-sponsored materials?

_____
_____
_____
_____

4. How do financial incentives affect medical training?

_____

_____

_____

_____

5. What can be done to promote unbiased education in medicine?

_____

_____

_____

_____

# Government and Regulatory Capture

**5 Insights:**

1. Big Pharma exerts significant influence over government agencies.
2. Regulatory capture can lead to lenient oversight of the pharmaceutical industry.
3. Conflicts of interest within agencies compromise drug safety and efficacy.
4. Policies may prioritize industry interests over public health.
5. Whistleblowers face retaliation for exposing misconduct.

**5 Reflection Questions:**

1. How does Big Pharma influence government agencies?

———————————————

———————————————

———————————————

2. What are the consequences of regulatory capture on drug safety?

_____

_____

_____

_____

3. How do conflicts of interest within agencies affect public trust?

_____

_____

_____

_____

4. Why might policies favor pharmaceutical companies over public health?

_____

5. What risks do whistleblowers face when exposing industry misconduct?

# Vaccine Controversies and Cover-Ups

**5 Insights:**

1. Vaccine safety controversies often involve suppressed data.
2. Some vaccines are linked to serious health issues.
3. Government and pharmaceutical companies may downplay risks.
4. Cover-ups can undermine public trust in vaccination programs.
5. Transparency is critical for addressing vaccine-related concerns.

**5 Reflection Questions:**

1. What data about vaccine safety is often suppressed?

_____

_____

_____

2. How can vaccines be linked to serious health issues?

_____

_____

_____

_____

3. Why might risks associated with vaccines be downplayed?

_____

_____

_____

_____

4. How do cover-ups affect public trust in vaccines?

_____

_____

_____

_____

5. What measures can improve transparency in vaccine safety?

_____

_____

_____

# Case Studies of Harm: Real Lives, Real Impact

**5 Insights:**

1. Real-life cases highlight the severe impact of pharmaceutical products.
2. Individuals harmed by drugs often face long-term health issues.
3. Legal and financial struggles accompany many harm cases.
4. Public awareness of these cases is often limited.
5. Personal stories reveal systemic failures in drug safety.

**5 Reflection Questions:**

1. What are some significant real-life cases of harm caused by pharmaceuticals?

_____

_____

_____

2. How do harmed individuals cope with long-term health issues?

_____

_____

_____

_____

3. What challenges do victims of pharmaceutical harm face?

_____

_____

_____

_____

4. Why is public awareness of drug-related harm often limited?

_____

_____

_____

_____

5. How do personal stories illustrate systemic failures in drug safety?

_____

_____

_____

_____

# Exposing Secretive Funding and Fake Independence

**5 Insights:**

1. Big Pharma often funds studies and nonprofits under the guise of independence.
2. Secretive funding can bias research outcomes and recommendations.
3. Independent studies may not be as unbiased as they appear.
4. Transparency in funding is crucial for credible research.
5. Hidden financial interests can mislead public health policies.

**5 Reflection Questions:**

1. How does secretive funding influence research outcomes?

_____

_____

_____

2. What are the signs of biased research due to hidden funding?

_____

_____

_____

_____

3. Why is transparency in funding important for research credibility?

_____

_____

_____

_____

4. How can hidden financial interests affect public health policies?

_____

5. What steps can be taken to uncover and address secretive funding?

# Ethics Violations and Professional Accountability

**5 Insights:**

1. Ethics violations in pharmaceuticals undermine public trust and safety.
2. Professionals involved in misconduct may evade accountability.
3. Ethical breaches often result in harm to patients.
4. Regulatory bodies may be slow to enforce ethical standards.
5. Strong accountability mechanisms are essential for industry integrity.

**5 Reflection Questions:**

1. What are common examples of ethics violations in the pharmaceutical industry?

_____

_____

_____

2. How can ethics violations impact patient safety?

   _____

   _____

   _____

   _____

3. Why might professionals evade accountability for misconduct?

   _____

   _____

   _____

   _____

4. What challenges do regulatory bodies face in enforcing ethical standards?

   _____

_____

_____

_____

5. How can accountability mechanisms be improved to ensure industry integrity?

_____

_____

_____

_____

# Tactics of Denial and Suppression

**5 Insights:**

1. Big Pharma uses denial to dismiss allegations and maintain a positive image.
2. Suppression tactics include silencing critics and controlling information.
3. Legal threats and smear campaigns are common methods to protect interests.
4. Suppression undermines public trust and transparency.
5. Exposing these tactics is crucial for accountability.

**5 Reflection Questions:**

1. How does Big Pharma use denial to handle allegations?

_____

_____

_____

_____

2. What methods are employed to suppress information about pharmaceutical misconduct?

_____

_____

_____

_____

3. How do legal threats and smear campaigns affect critics?

_____

_____

_____

_____

4. Why is suppression detrimental to public trust?

_____

_____

_____

_____

5. What strategies can be used to expose and counteract these tactics?

_____

_____

_____

_____

# A Call to Action: Advocacy, Awareness, and Reform

**5 Insights:**

1. Advocacy is essential for promoting transparency and ethical practices.
2. Raising awareness can mobilize public support for reform.
3. Effective reform requires collaboration among various stakeholders.
4. Grassroots efforts can drive significant change in healthcare.
5. Continuous vigilance is needed to ensure lasting improvements.

**5 Reflection Questions:**

1. How can advocacy promote transparency in the pharmaceutical industry?

_____

_____

_____

_____

2. What methods are effective in raising public awareness about healthcare issues?

_____

_____

_____

_____

3. Why is collaboration crucial for successful reform?

_____

_____

_____

_____

4. How can grassroots efforts influence healthcare policies?

_____

5. What steps are necessary to maintain long-term improvements in the industry?

www.ingramcontent.com/pod-product-compliance
Lightning Source LLC
Chambersburg PA
CBHW070955220526
45471CB00007B/3045